THE SLOWPOKE SNAIL · THE SLOWPOKE SNAIL

如果你想听我的故事，就需要有一点耐心，不然的话，你会连我真正的样子都看不清。

If you want to hear my story, you will need a little patience; otherwise, you will not even be able to see what I look like.

······ 蜗 牛 慢 吞 吞 ······The Slowpoke Snail······

······ 撰文　周宗伟　Text by Zhou Zongwei ······ 绘图　朱赢椿　Illustrations by Zhu Yingchun ······

·2·3· 我走得很慢很慢，
I plod along as slow as slow can be,
并不是因为我懒散，
Not because I am lazy,
而是我天生一副柔弱微小的身体，
But because I was born with this tiny frail physique,
只能背着壳缓缓地爬行。
I can only crawl along slowly, bearing this shell on my back.

虽然我走得很慢，

Although I walk very slowly

但走路已变成了习惯。

I have gotten into the habit of walking.

走路本身机械又枯燥，

Walking itself is mechanical and dull,

可停下来反而更觉无聊。

Yet I feel even more bored standing still.

无论路上的风景如何绚烂，

No matter how splendid the roadside scenery is,

我的内心却总是茫然。

In my heart I feel at a loss.

祖先教我们缓慢地生活，

Ancestors taught us to live at a slow leisurely pace,

我从未对此产生怀疑，

I have never doubted about this

直至一天在路上，

Until one day on the road

看到了被踩扁的同伴，

I saw a companion get flattened underfoot.

他颤抖的样子让我触目惊心，

The way he quivered shocked me dreadfully,

行人的脚步让他丧生，

Passing footsteps ended his life,

"慢"却救不了他的命。

And his "slowness" could not save him.

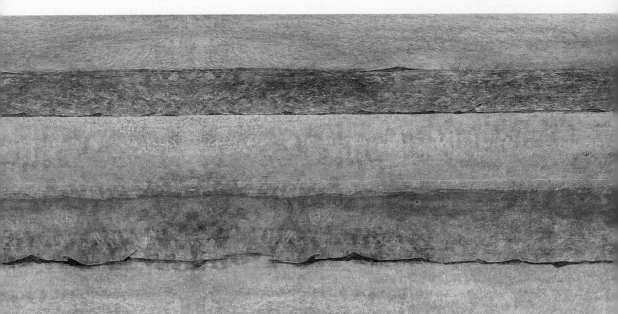

对死亡的恐惧,
Due to fear of death,

让我认真思考"活着"的问题,
I earnestly pondered the problem of "living".

怎样活着才有意义?
How can living be meaningful?

既然"慢"不能让我们活得更好,
Since "slowness" cannot make my life better,

那么"快"就该是真理。
Then the truth must lie in being "fast".

我想学习﹃快﹄的生活，

I wanted to learn about living "fast".

随处可见比我﹃快﹄的老师——蚂蚁、瓢虫，还有毛毛虫，

In many places I saw teachers faster than me—such as ants and ladybugs, And also caterpillars.

当然，

Of course

更快的还是黄蜂。

The wasp was even faster.

我虚心地向黄蜂请教﹃快﹄的问题，

I humbly asked for instruction in being fast.

他热心地给我示范如何快速地飞行。

He enthusiastically showed me how fast he could fly.

我惊叹于他高超的飞行技巧，
I was amazed by his accomplishments in flying.

却见他飞得太快而不小心撞上了蛛网，
But he went too fast and carelessly flew into a spider web.

片刻间一命归西。
In a moment's time he gave up the ghost.

我替他惋惜又暗自庆幸，
I felt terrible for him, yet I rejoiced to myself.

慢的好处真要细心才能领会，
It takes careful thought to grasp the benefit of slowness.

祖先的智慧自有道理。
Our ancestors' wisdom makes sense after all.

我继续缓缓地、缓缓地爬行。

I continued crawling in my slow laggardly way;

路边传来一阵呻吟，

From the roadside I could hear a moaning sound.

是一只年老的蜗牛摔破了壳，

It was an old snail whose shell had been broken in a fall.

正被蚂蚁们啮咬着身体。

There were ants gnawing at her flesh.

可怜她已经老得走不动路，

Pitiful old soul——she was no longer able to walk.

只能任由欺凌。

She could only submit to her torments.

我上前安慰她：放宽心，一切都会好的。

I approached and spoke reassuring words. "Put your mind at ease. Things will work out in the end."

她无奈地摇头道：没有用的，等你自己到了这一天就会知道。

She shook her head in resignation. "It is no use. When your dark day comes, you will know."

我们已习惯于用安慰去表示关心，

It was our habit to say comforting words, to show concern.

明知安慰只是欺人并自欺，

We knew our reassurances were deceptive to both of us.

却没有面对真相的勇气。

But we lacked courage to face the truth.

看着老蜗牛已奄奄一息，
Seeing the old snail near her last gasp,

我虽然义愤填膺，
I was filled with righteous indignation,

却无能为力，
But I was powerless to help.

只恨我们慢得竟然躲不过小小的蚂蚁。
I could only rue our slowness that makes us prey even to ants.

这才知道，
Then I realized;

慢本身并不是过错，
Being slow is not wrong in itself;

错的是因为缓慢而滋生的懦弱性格。
What's wrong is the cowardice that slowness fosters in us.

我发誓要做一只坚强而勇敢的蜗牛，

I vowed to be a brave and steadfast snail;

绝不向强势低头。

Never would I lower my head before brute force.

刚下定决心，

No sooner had I made this resolution,

就遇到了险境。

Than I encountered a dire predicament.

只见一只怪物趴在我的面前，

I saw a monster squatting right in front of me.

张牙舞爪地露出挑衅的恶脸。

It struck a fighting pose and leered combatively.

我把头一昂，

I held my head high,

想先下手为强。

Intending to strike first and get the upper hand.

我冲上去和他扭打，
I charged forward and grappled with him.

混战中我们一起摔倒在地。
In the melee we both fell to the ground.

他带着尖刺的爪子割破了我柔嫩的身体，
His sharp claws dug into my soft underbelly.

一股剧痛袭来，
A sharp pain shot through me;

我一阵眩晕。
I fainted away.

当我从疼痛中醒来，

When the pain passed and I revived,

发现他也一动不动地躺在那里，

I saw that he was lying there motionless too

仿佛已经死去。

As if he were dead.

我定神细看，

I took a careful look,

才发现，

Only to discover

原来那只是一个蝉蜕而已，

It was just a cicada's carapace.

我顿时呆若木鸡。

I was dumbfounded.

我在伤痛中冷静反省，ㆍ26ㆍ27ㆍ
Despite the pain I calmed myself and reflected;
自己竟然荒唐地向一个毫无生命的空壳搏击，
How absurd I had been, to strike at a lifeless shell.
它根本不会伤害我，
There was no way it could have hurt me;
一切都是自己在伤害自己，
I had brought it all on myself.
我们常常是自己给自己树敌。
We often make enemies for ourselves.

我拖着伤痕累累的身体，
Dragging my scarred body

更加缓慢地前行，
I went onward even more slowly,

心中昂扬的斗志也没了踪影。
My haughty fighting spirit was gone without a trace.

路边躺着一具锹甲的尸体，
Beside the road there lay the body of a scarab

他巨大的身躯即便在死后也散发着强大的威慑力，
Even in death, his hulking corpse had an intimidating air.

让许多小虫不寒而栗。
Many little insects shuddered at the sight.

我此刻却没有丝毫的畏惧，

Right then I did not have the slightest fear;

反而对他生出了一些怜悯。

I even felt a touch of pity for him,

想他生前雄霸的身躯曾经何等风光，

Thinking what an impressive figure he had cut while alive,

如今却也落得被虫蚁啃噬的下场。

But now he was reduced to being gnawed on by ants.

原来外表的凶悍并不是真正的勇敢和坚强。

In the end surface ferocity is not true courage and strength.

我有些累了，

I felt a bit tired.

藏到一片叶子下面休息。

So I hid under a leaf to rest.

空中飘下一阵浓雾，

A cloud of thick mist floated down from the sky,

夹着一股强烈的农药气息。

Bringing a thick smell of pesticide.

我马上警惕，

I was on guard immediately,

准备躲进壳里。

Ready to pull back into my shell.

却听见地上传来呼救的声音，

I could hear cries for help, coming from close to the ground.

叶子下露出一只西瓜虫挣扎的身影。

A pill-bug was thrashing about under a leaf.

西瓜虫仰面朝天地躺在地上，

The pill—bug was lying supine on the ground.

挥舞着细腿翻来翻去。

Waving its tiny legs as it rocked back and forth.

农药熏得他失去了力气，

The pesticide's toxic effect had robbed him of strength.

任凭他怎么努力就是翻不过身体。

However he tried, he could not turn upright.

我伸出一只触角拉了他一把，

I reached out a feeler and gave him a pull;

他终于翻过身来，

At last he righted himself;

对我感激不尽。

He was quite grateful to me.

西瓜虫和我成了朋友，

The pill—bug and I became friends；

常常陪我一起散步，

He often joined me on my walks．

从此，我走路不再孤独。

From then on, my walks were not lonely．

那些农药让许多昆虫丧生，

That pesticide cloud took the lives of many insects．

幸存让我们懂得了珍惜和感恩。

By surviving, we learned to cherish life and be grateful．

尽管西瓜虫对我很有耐心，
Although the pill—bug was patient with me

为了我故意走得很慢，
And deliberately slowed his pace for me,

可我还是为自己的缓慢感到不安。
Being so laggardly made me feel ashamed.

为了让我宽心，
In order to put my mind at ease,

西瓜虫又带我结识了一个新的伙伴——尺蠖，
The pill—bug helped me to make a new friend——an inchworm,

他弓着腰走路的样子真是好玩。
His bent—waisted way of walking was quite laughable.

尺蠖真的很风趣，

The inchworm was a witty fellow.

边向我展示他一拱一拱走路的姿势，

As he demonstrated his arch—backed walking posture

边对我调侃道：看，我不仅走得也很慢，

He joked with me, saying:Look, my walk is not only slow,

而且走得还很难看，

It is also unattractive,

可是这和我们的友谊无关。

But that has nothing to do with our friendship.

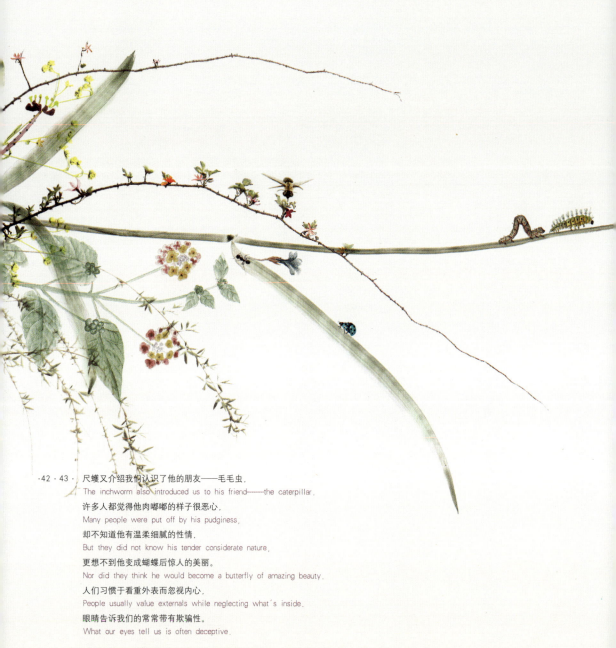

尺蠖又介绍我们认识了他的朋友——毛毛虫，
The inchworm also introduced us to his friend——the caterpillar.

许多人都觉得他肉嘟嘟的样子很恶心，
Many people were put off by his pudginess,

却不知道他有温柔细腻的性情，
But they did not know his tender considerate nature,

更想不到他变成蝴蝶后惊人的美丽。
Nor did they think he would become a butterfly of amazing beauty.

人们习惯于看重外表而忽视内心，
People usually value externals while neglecting what's inside.

眼睛告诉我们的常常带有欺骗性。
What our eyes tell us is often deceptive.

我们几个朋友从此形影不离，
From then on the four of us were inseparable.

终日结伴一同嬉戏，
We spent whole days frolicking in each other's company.

快乐无比。
Our happiness was incomparable.

我以为找到了"活着"的意义，
I supposed I had found what it meant to "live".

也体会到只有"爱"才能让自己真正强大，
I realized that only love can make a person strong.

心中对死亡的恐惧渐渐淡去。
My fear of death gradually diminished.

朋友们总是渴望永久的相聚,
Friends always hope their togetherness can last forever,

大家彼此相互依赖,
They want to keep supporting each other,

紧密联结成一体,
And to be bound together as one body.

这种感觉很甜蜜。
This is a sweet feeling. The regrettable thing is

遗憾的是,天下没有不散的筵席,
In this world all revels must break up sometime;

甜蜜也未必是好东西。
Perhaps sweetness is not the best feeling to have;

快乐到了极致,危险就在一旁紧伺。
When happiness reaches an extreme danger awaits around the corner.

危急之中来不及思索，
When danger strikes there is no time to think over

螳螂逼得我们从树梢跌落，
A preying mantis attacked, making us fall from a high branch.

一阵茫然失措……
We blanked out for a spell...

醒来时，

When I revived,

我发现自己漂在水里，

I found that I was floating on water.

身边浮荡着朋友们的尸体。

Bobbing alongside of me were corpses of my friends.

我借助于祖先的馈赠，

My ancestral gift had proved helpful;

因为外壳的浮力，

Because the buoyancy of my shell

而保全了性命，

My life had been preserved,

可我内心却悲伤无比。

But nothing could have been worse than my sorrow.

我随风漂到了岸边，

I was tossed by the wind onto a stream bank.

虽然死里逃生，

Though come through alive,

却感到生活兴味索然。

I had lost enthusiasm for life.

朋友们的死让我失魂落魄，

The death of my friends left me dispirited.

我背着壳仿佛背着一座大山，

Carrying my shell was like bearing a huge mountain.

脚步凝重，

My steps were halting;

步履越发缓慢。

My pace grew more and more laggardly.

我独自在岸边徘徊，

I paced alone on the stream bank,

总以为还能在水中看到朋友们的身影。

Thinking I might see my friends' shapes in the water.

恍惚间看见一只水蛤蚜扑过来撕咬我的身体，

Then a water strider pounced and clamped jaws on my body.

吓得我拼命挣扎躲闪，

In great fright I struggled to break free,

无济于事，

But it did no good;

只得大呼救命。

All I could do was cry out for help.

一只蜻蜓循声前来，

A dragonfly came to see what the noise was about,

诧异地问我遇何危险。

He quizzically asked what danger I had met with,

我说是水蚂蚱在咬我，

I said a water strider was biting me,

蜻蜓哑然失笑道：

The dragonfly let out a chuckle and said,

"水蚂蚱在水面上，

"The water strider is on the water's surface;

咬的只是你的影子而已。"

He is only biting your reflection,"

我好似梦醒，

I felt as if waking from a dream,

回神一看，

Collecting my wits I looked closely;

才发现自己竟然错把虚幻的倒影当成了自己。

I had mistaken an illusory reflection for myself,

我清醒过来后惭愧不已，

Upon sober reflection I was quite embarrassed.

自己过度沉溺于痛苦以致迷失了本性，

Being immersed in pain, I had lost sight of my true nature

竟然到了真假不分的境地。

To the point that I confused truth with illusion.

蜻蜓却笑着安慰道：

But the dragonfly comforted me：

不必自责，

No need to blame yourself.

人生亦不过是一场巨大的梦境，

Life is nothing but one big dream.

有几个真的清醒？

How many of us are truly awake？

蜻蜓的话如醍醐灌顶，

Listening to him was like an anointment with balm,

为了摆脱痛失爱友的低落情绪，

I wanted to break away from my grief-stricken mood

我打算换换环境，

So I planned on a change of scene,

跟随蜻蜓离开伤心的水岸，

Following the dragonfly, I left that dismal stream bank

来到一片新天地。

And found a place under new skies,

蜻蜓的智慧让我钦佩，

I admired the dragonfly for his wisdom,

我又获得了新的友谊。

Once again I found friendship,

盛夏来临，酷暑难当。

The height of summer was upon us; the sweltering heat was hard to endure.

连日无雨，旱情危急。

Day after day no rain fell; drought were critical.

我再次得益于祖先的馈赠。

Once again my ancestral gift proved a blessing.

躲进壳里避暑，

I withdrew into my shell to escape the dryness,

蜻蜓的生活却日渐艰难。

But the dragonfly's life grew tougher every day.

可是，他却坦然道：

Even so, he stoically said to me:

活着的目的并不是为了体验恐惧，

To experience dread is not the point of life.

只需让该来的来，

We only have to let what comes come

让该去的去。

And let go of what must go.

炎炎烈日也许会晒伤蜻蜓的翅膀。

The blazing sun was liable to scorch the dragonfly's wings.

我日渐担忧蜻蜓的安危,

Each day I grew more worried for his safety.

劝他快些找个安全的地方躲藏,

I urged him to find a safe hiding place.

他却宽慰我道:

But he put my mind at ease, saying:

人人都喜欢计划,

Everyone likes to make plans;

却不知计划永远赶不上变化,

They don't know that plans never catch up with reality.

因为世事本无常。

Because things of this world are impermanent.

既然斗不过无常,

Since we cannot overcome impermanence,

不如珍惜当下。

It is better to cherish the present moment.

好不容易盼来了一阵雨

我从壳里伸出头来呼吸新鲜空气.

却看不见蜻蜓的影子.

我心中有了种不祥的预感.

彷徨又孤单.

虽然我也懂得生活要"随遇而安",

可说起来容易.

做起来真的很难.

我终于找到了蜻蜓，

I finally found the dragonfly,

可是他已经停止了呼吸。

But he had already stopped breathing.

他的样子安详而宁静，

His appearance was serene and peaceful,

仿佛只是在休息。

As if he were resting.

他在死亡面前一定没有恐惧，

I was sure he had faced death without dread,

因为他曾经说过：

Because, as he had once said,

死亡也只是一种幻境。

Death is just another illusory state.

人人都畏惧死亡，

Everyone is afraid of death,

却不知正是死亡才能教会我们要懂得舍弃。

Little do they know that death can teach us renunciation.

失去朋友的痛苦让我再次警醒：

Once more the pain of loss roused my awareness;

无常不能逃避，

There is no escaping impermanence;

不能抵抗，

It cannot be resisted;

它不是一个敌人，

It is not an enemy;

无法用斗争去消灭它。

There is no way to vanquish it by struggle.

之所以会害怕无常，

The reason we fear impermanence

是因为我们不愿随顺而总想抵抗。

Is that we refuse to follow along but wish to resist it.

一旦懂得了随顺，

Once we have learned to go along with it,

学会了接纳，

Have learned to accept it,

无常便不再是无常，

Impermanence is no longer impermanence;

一切只不过是平常。

Whatever happens is simply ordinary.

我怀着对蜻蜓的祝福，
With blessings in my heart for the dragonfly,
尝试独立的生活，
I tried to live on my own.
想起他的鼓舞，
When lonely, I thought of his encouragement;
孤单时再也不会顾影自怜。
This kept me from slipping into narcissism.
我依旧缓慢地走路，
As always I walked in my slowpoke way,
开始细心地品味路上的风景。
I began taking pains to savor the roadside scenery.
因为放慢脚步，
Due to my leisurely pace,
才感受到了身边存在的美好事物，
I could perceive the wonders around me,
生活也因此平添了许多乐趣。
Because of this, life brought additional joys.

一日，
One day

在路上邂逅了一只黑色的蜗牛，
Along the way I encountered a black snail.

我高兴地上前打招呼，
I gladly went forward to greet him,

想和他交朋友。
Thinking we could be friends.

他却嫌自己长得太丑，
But he was convinced of his own ugliness;

而自惭形秽，
He held himself to be physically repulsive.

在我面前自卑得抬不起头。
He lacked the self—esteem to raise his head before me.

我开玩笑地说：
I jokingly said,

蜗牛的世界中可没有种族歧视，
"In the world of snails there is no racial prejudice.

其实你一点也不丑。
Actually you are not ugly at all."

我的真诚让黑蜗牛消除了戒心，

My earnestness helped to eliminate the black snail's inferiority.

愿意随我同行。

He was willing to go walking with me.

走着走着，

As we walked along,

却遇到一大团泥巴横在路中央，

We came upon a large clump of mud lying in our path.

把路堵了个正着，

Our way forward was blocked.

我只好用身子去拱开它。

I had to press against it, to shove it aside.

泥巴自己却抖动起来，

Then the mud itself started quivering;

下面颤巍巍地伸出一对触角，

Out from underneath poked a pair of feelers.

它竟然也是一只蜗牛。

It turned to be a snail like us.

黑蜗牛仿佛受到了安慰，

The black snail seemed to take comfort from this.

忍不住笑道：

He could not help but say,

没想到世上居然还有比我更丑的，

Who knew there is something uglier than me in this world?

看来我还不算糟糕。

My situation is not such a mess after all.

泥蜗牛正羞愧难当，又想缩回壳里。

The mud snail was terribly humiliated, It wanted to pull back into its shell.

恰逢一只路过的蛞蝓听了我们的对话，

Just then a passing slug heard our conversation.

前来开导他：

It drew near and tried to make him see reason:

"你以为你丑，我岂不比你更丑？

"You think you are ugly, but aren't I even uglier than you?

人们因为我的丑陋还给我起了外号叫'鼻涕虫'，

Due to my ugliness people have nicknamed me 'snot worm,'

可笑的是别人，

But they are the laughable ones.

我并不会损失一分一毫。

What they say doesn't diminish me in the slightest.

平凡的外表恰恰是最安全的庇护。

A plain exterior is the best way to ensure safety.

人们喜欢漂亮出众，

People like to be attractive, to stand out.

却不知其实平常才是福。"

They forget that true good fortune lies in being ordinary."

蛞蝓的话让大家如饮甘露。

The slug's words were like soothing balm for us.

我们原本是同类，又何必用外表的差异拉大内心的距离？

We were originally kindred, Why should we use outer differences to magnify inner distance？

不必总拿自己和别人作比较。

We need not always compare ourselves with others.

比较有时能帮助自己进步，有时却是一种无知的糊涂。

Sometimes comparison can help you make progress, But sometimes it is just ignorance and confusion.

经历得越多，我的心境也越平和，
The more experience I had, the more calm and steady my mind became,

快也好，慢也好，美也好，丑也好，热闹也好，冷清也好，
Whether fast or slow, beautiful or ugly, Whether bustling or deserted,

只要心里放平，
As long as you keep a steady mind,

怎样"活着"其实都好。
"live" in any way will be alright,

"活着"本身就是最大的财富。
"Living" itself is the greatest wealth,

我愈加珍惜"活着"的每一刻，
I cherished each instant of "living" all the more,

认真地走好缓慢的每一步。
And took pains to make each slow step a good one,

有趣的是，

The interesting thing was,

当我不再害怕孤单，

When I no longer feared loneliness,

朋友反而多了起来。

Strangely enough, I started making more friends.

原来孤单者只是自己在排斥自己，

It turns out that lonely people are only ostracizing themselves.

当他们关闭了自己的心门，

When they close the doors of their heart,

别人走不进来，

Other people cannot get in,

自己也永远走不出去。

And they themselves can never get out.

我和朋友们玩一种"叠罗汉"的游戏，
My friends and I played a game called "stacking arhats",
这种高难度的动作需要彼此亲密合作，
Such a difficult feat required close cooperation among us.
只有心无芥蒂的关系才能创造高度的默契。
Only friendships without grievance allow tacit understanding.
即便在玩耍中，
Even at play,
信任也是快乐的前提。
Trust is a precondition of happiness.

在高处玩耍正得意忘形的时候，
While thoughtlessly indulging in acrobatic games,

我一不留心栽了跟头，
In a careless moment I fell head over heels,

跌破了外壳。
My shell was broken in the fall,

这才感觉我们终日背负的这副外壳真是一个累赘，
I started to feel these shells we carry are an encumbrance,

为什么不能像蛞蝓一样，
Why can't we be like those slugs,

不背包袱，
With no load to carry on their backs?

多么轻松和潇洒？
How carefree that would be!

疼痛中想起老蜗牛曾说过的话，
Being in pain, I thought of what the old snail had said,

越发体会深刻，
More than ever I realized,

痛苦长在别人身上总比在自己身上要轻得多。
Pain afflicting others carries far less weight than our own pain,

几只蚂蚁见了， ·92·93·
A few ants caught sight of me

垂涎欲滴地尾随上来，
And droolingly came after me.

咬我的伤口。
They sank their pincers into my wound.

我忍着痛努力向前爬，
I braced myself against the pain and crawled onward,

想躲过蚂蚁的追袭，
Trying my hardest to evade the pursuing ants,

可我这慢吞吞的速度，
But with a slowpoke pace like mine,

要想躲避蚂蚁无异于天方夜谭。
Expecting to get away from ants would be a fantasy.

幸亏遇见了一只好心的蜘蛛，

Luckily I came across a well—meaning spider,

他仗义相助，

For the sake of justice he came to my aid

在我的壳上织起了网。

By spinning a web over my shell,

蚂蚁害怕蛛网，

Being afraid of spider webs,

才悻悻地离去。

The ants left in chagrin.

我在蜘蛛的庇护下静静地养伤，

Under the spider's protection, I convalesced quietly.

伤痛之中不禁感叹道：

In my pain I could not help but exclaim:

"我们蜗牛祖祖辈辈都背着这副外壳生活，

"Through the ages, we have lived with these shells on our backs,

以为它能给我们带来幸福，

Supposing they bring us good fortune,

可有时候它也真的给我们带来痛苦。

But sometimes they cause us grief.

明知它是个负担，

Though we know they are burdensome,

可放下它也真的很难！"

Still it is hard to let them go!"

蜘蛛听了，劝导我道：

The spider offered words of advice:

"你们蜗牛害人之心虽没有，

"You snails have no evil intentions toward others,

但这防人之壳却也不可无。

But never forget to guard against others'

真正应当放下的不是背上的壳，

The shell on your back is not what you should let go of,

而是心上的重负。"

Rather, it is the burden you carry in your mind."

蜘蛛之语让我豁然开朗，

These words suddenly enlightened me.

等养好了伤，

After recovering from injury,

我告别了蜘蛛，

I said farewell to the spider

又上了路。

And got back on the road.

天气越来越热,

The weather grew increasingly hot.

我正一路寻找阴凉,

Along the way I looked for a shady spot.

却撞上了一队搬家的蚂蚁。

But I ran into a troop of migrant ants.

想起曾经遭受他们的欺负,

Remembering the mistreatment I had suffered,

不由得心头火起,

My heart burned with fury,

但此刻他们对我并无恶意。

But this time they had no ill intentions.

冷静思忖,

Thinking it over calmly,

才更觉得"把心放平"谈何容易?

I felt that "keeping a steady mind" was no easy matter.

世上最难征服的就是自己的心。

In this world the hardest thing to conquer is one's own mind.

我努力说服自己，

I tried hard to persuade myself

不与蚂蚁们为敌。

Not to treat the ants as enemies.

想想他们的一切作为都是为了生存，

Considering that everything they did was for survival

便理解和宽恕了他们。

I got to understand and forgive them.

心中没有了仇恨，

Once my heart was free of hatred,

立刻感到了轻松和解脱，

I soon felt at ease and liberated.

原来一切仇恨只是捆绑自己的无形枷锁。

Hateful thoughts are like invisible fetters that bind us.

我对蚂蚁们的友好态度得到了回应,

My friendliness to the ants drew out a response,

他们也好心地提醒我:

They considerately reminded me

很快会有暴雨来临,

That a thunderstorm would come soon,

低矮的地方不安全,

Low places would not be safe;

必须赶快向高处转移。

It was necessary to move to higher ground.

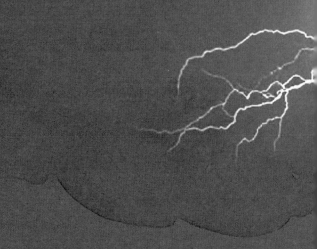

蚂蚁的预言果然灵验。
Sure enough, the ants' prediction came true.

没过多久，
Before long

天空电闪雷鸣，
The sky flashed with lightning and rumbled with thunder

一场从未见过的暴雨自天而降，
A cloudburst I had never seen poured down,

所幸的是，
Luckily for me

我已在高处找到了一个比较安全的地方。
I had already found a safe place on higher ground.

瓢泼大雨下个不停，

The rain poured down in sheets, on and on.

夜晚随后降临，

And then came nightfall.

天空一片黑暗。

The sky was enveloped in darkness.

除了哗哗的雨声在耳边作响，

The sigh of rain sounded in my ears,

什么也看不见，

But I could not see anything.

我第一次感觉到死亡的距离原来是这样的近。

For the first time I really felt the nearness of death.

仿佛等了一个世纪，

After what seemed like a century of waiting,

暴雨终于停息，

The storm finally stopped.

天空出现几点飘忽的星光，

Specks of starlight seemed to float in the heavens.

在雨后的夜空格外美丽。

After a rainfall, the night sky was especially beautiful.

我正为这景色陶醉痴迷，

I was entranced and enraptured by the scene.

怎知这美丽之中却暗藏了杀机。

Could deathly threats lurk even in such beauty?

原来那是萤火虫的光芒，

The stars turned out to be fireflies.

他们是蜗牛的天敌。

They are natural enemy of snail.

此时，忍耐是生存的唯一之计，

At such times, patience is the key to survival.

我只得屏住呼吸一动不动，

I could only stay put, with bated breath.

静静等待他们离去。

And quietly wait until they left.

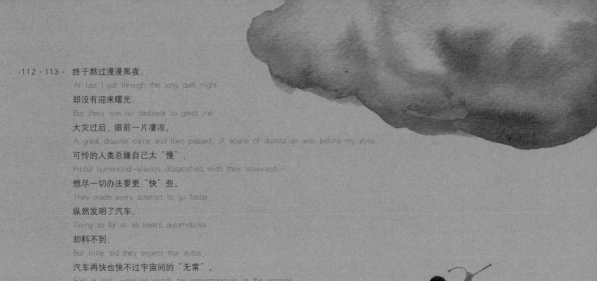

终于熬过漫漫黑夜，

At last I got through the long dark night,

却没有迎来曙光，

But there was no daybreak to greet me.

大灾过后，眼前一片凄凉。

A great disaster come and then passed. A scene of desolation was before my eyes.

可怜的人类总嫌自己太"慢"，

Pitiful humankind—always dissatisfied with their slowness—

想尽一切办法要更"快"些，

They made every attempt to go faster.

纵然发明了汽车，

Going so far as to invent automobiles,

却料不到，

But little did they expect that autos,

汽车再快也快不过宇宙间的"无常"。

Fast or not, were no match for impermanence in the cosmos.

我活了下来，
I made it through alive.

那些跑得快的生命却未能幸免。
But those fast--moving lives could not escape their destiny.

在灾难面前，
In the face of disaster,

他们竟然不如一只慢吞吞的蜗牛，
They were no match for a slowpoke snail.

大自然的这个玩笑似乎开得大了一点。
This joke played by nature seems a bit heavy handed.

大水退去，
The floodwaters receded

我又回到了生我养我的大地上，
I returned to the good earth that bore and nurtured me.

没有伙伴，
I had no companions;

没有食物，
I had nothing to eat;

我独自一个，
I was all by myself,

一无所有，
And had nothing to call my own.

缓缓地、缓缓地爬行。
I went on crawling in my laggardly slowpoke way.

我虽然不知道未来该向何方，

Although I do not know my future direction，

却知道了——真正的 "爱"，

I have learned that true "love"

是当你孤独到极点时，

Is when you reach the depths of loneliness

仍旧对世界心怀感激；

Yet still hold gratitude toward this world in your heart，

当你一无所有时，

And when you have nothing to call your own，

仍旧对天地万物心存善意。

Yet still think kindly of all things under heaven.

平安, ·120·121·
Being peaceful and safe
就是幸福。
Is blessedness.
我继续慢吞吞地走路,
I keep on walking in my slowpoke way
自在而满足。
At ease and satisfied.

作者简介
The Information of Author

周宗伟 博士

南京师范大学心理学院副教授

主要从事艺术治疗、心理健康教育等相关领域的实践与教学研究工作，是国家二级心理咨询师，美国 NGH 催眠协会注册催眠师。著有《高贵与卑贱的距离——学校文化的社会学研究》。

Zongwei ZHOU, PhD

Associate professor, School of Psychology, Nanjing Normal University

Her recent research has focused on art therapy, mental health education and other related fields. She is a National Psychological Consultant (Level 2) and NGH Certified Hypnotherapist. And the author of *The Distance Between Noble and Humble: A Sociological Study of School Culture.*

朱赢椿

南京师范大学书文化研究中心主任，全国新闻出版领军人才

由他创作与设计的图书数次被评为"中国最美的书"和"世界最美的书"。2008 年，《蚁呓》被联合国教科文组织德国委员会评为年度"最美图书特别制作奖"。2017 年，《虫子书》荣获年度"世界最美的书"银奖，并被大英图书馆永久收藏。

以《虫子书》为代表，《虫子旁》《虫子本》《虫子诗》相继出版——微观世界，蔚为大观。同时，"虫子说"系列也逐渐走向世界，《蚁呓》《蜗牛慢吞吞》《蛛嘱》先后版权输出至韩国、捷克等国。2022 年，《蜗牛慢吞吞》还被改编为舞台剧在德国上演。

理想国还出版有朱赢椿作品《设计诗》《语录杜尚》《便形鸟》。

Yingchun ZHU

Director of Book Culture Research Center, Nanjing Normal University

National Press and Publication Leading Talents

His books and book design works have won and been nominated 'Beauty of Books in China' Award and 'Best Book Design from all over the World' (Stiftung Buchkunst) many times. In 2008, the UNESCO German Commission identified *ANT* as that year's 'Most Beautiful Book'. In 2017, *THE LANGUAGE OF BUGS* won the Silver Medal Winner in'Best Book Design from all over the World', and was in the permanent collection of the British Library.

To be by bugs side, Yingchun ZHU created a BUG UNIVERSE by a series of bug-oriented books.

Beijing Imaginist Time also published other works by Yingchun ZHU such as *POETRY IN DESIGNED,QUOTATIONS FROM DUCHAMP,CACAFORM BIRDS.*

图书在版编目(CIP)数据

蜗牛慢吞吞 / 周宗伟著；朱赢椿绘.
—桂林：广西师范大学出版社，2011.9（2022.6重印）
ISBN 978-7-5495-0216-5

Ⅰ.①蜗… Ⅱ.①周…②朱… Ⅲ.①人生哲学－通俗读物
Ⅳ.①B821-49

中国版本图书馆CIP数据核字(2010)第226428号

广西师范大学出版社出版发行

广西桂林市五里店路9号　邮政编码：541004
网址：www.bbtpress.com

出　版　人：黄轩庄
责任编辑：杨静武
装帧设计：朱赢椿　皇甫珊珊
全国新华书店经销
发行热线：010-64284815
北京华联印刷有限公司印刷

开本：889mm×1194mm　1/24
印张：6　字数：10千字　图片：65幅
2011年9月第1版　2022年6月第7次印刷
定价：69.00元（精装）

如发现印装质量问题，影响阅读，请与出版社发行部门联系调换。

朱赢椿创作《蜗牛慢吞吞》的工作台
Zhu Yingchun's worktable, where this book was created

朱赢椿饲养的蜗牛
Zhu's pet snail

漫步和熟睡的蜗牛
Snails in stroll and sound sleep

书里的许多情节在现实中发生——被踩扁的蜗牛
Many scenes from this book happened in real life:
A snail flattened underfoot

——蜘蛛在破了的壳上结网
A spider span a web on a broken shell

——叠罗汉
Stacking arhats

THE SLOWPOKE SNAIL · THE SLOWPOKE SNAIL